GOD'S DESIGN® FOR THE PHYSICAL WORLD

INVENTIONS & TECHNOLOGY

TEACHER SUPPLEMENT

1:1
answersingenesis
Petersburg, Kentucky, USA

ANSWERS IN GENESIS SCIENCE BY DEBBIE & RICHARD LAWRENCE

God's Design for the Physical World
Inventions & Technology Teacher Supplement

Second printing: April 2012

Published by Answers in Genesis, 2800 Bullittsburg Church Rd., Petersburg KY 41080

You may contact the authors at (970) 686-5744.

ISBN: 1-60092-290-2

Cover design & layout: Diane King
Editors: Lori Jaworski, Gary Vaterlaus

Printed in China.

www.answersingenesis.org www.godsdesignscience.com

Table of Contents

Welcome to God's Design® for the Physical World

God's Design for the Physical World is a series that has been designed for use in teaching physical science to elementary and middle school students. It is divided into three books: *Heat and Energy, Machines and Motion,* and *Inventions and Technology.* Each book has 35 lessons including a final project that ties all of the lessons together.

In addition to the lessons, special features in each book include biographical information on interesting people as well as fun facts to make the subject more fun.

Although this is a complete curriculum, the information included here is just a beginning, so please feel free to add to each lesson as you see fit. A resource guide is included in the appendices to help you find additional information and resources. A list of supplies needed is included at the beginning of each lesson, while a master list of all supplies needed for the entire series can be found in the appendices.

Answer keys for all review questions, worksheets, quizzes, and the final exam are included here. Reproducible student worksheets and tests may be found on the supplementary CD-Rom for easy printing. Please contact Answers in Genesis if you wish to purchase a printed version of all the student materials, or go to www.AnswersBookstore.com.

If you wish to get through the three books of the *Physical World* series in one year, you should plan on covering approximately three lessons per week. The time required for each lesson varies depending on how much additional information you want to include, but you can plan on about 45 minutes per lesson.

If you wish to cover the material in more depth, you may add additional information and take a longer period of time to cover all the material or you could choose to do only one or two of the books in the series as a unit study.

Why Teach Physical Science?

Maybe you hate science or you just hate teaching it. Maybe you love science but don't quite know how to teach it to your children. Maybe science just doesn't seem as important as some of those other subjects you need to teach. Maybe you need a little motivation. If any of these descriptions fits you, then please consider the following.

It is not uncommon to question the need to teach your kids hands-on science in elementary school. We could argue that the knowledge gained in science will be needed later in life in order for your children to be more productive and well-rounded adults. We could argue that teaching your children science also teaches them logical and inductive thinking and reasoning skills, which are tools they will need to be more successful. We could argue that science is a necessity in this technological world in which we live. While all of these arguments are true, not one of them is the real reason that we should teach our children science. The most important reason to teach science in elementary school is to give your children an understanding that God is our Creator, and the Bible can be trusted. Teaching science from a creation perspective is one of the best ways to reinforce your children's faith in God and to help them counter the evolutionary propaganda they face every day.

God is the Master Creator of everything. His handiwork is all around us. Our Great Creator put in place all of the laws of physics, biology, and chemistry. These laws were put here for us to see His wisdom and power. In science, we see the hand of God at work more than in any other subject. Romans 1:20 says, "For since the creation of the world His invisible attributes are clearly seen, being understood by the things that are made, even His eternal power and Godhead, so that they [men] are without excuse." We need to help our children see God as Creator of the world around them so they will be able to recognize God and follow Him.

The study of physical science helps us to understand and appreciate the amazing way everything God created works together. The study of energy helps us understand that God set up the universe with enough energy to sustain life and that He created the sun to replenish the energy used up each day. The study of friction and movement helps us appreciate the laws of motion and helps us understand how simple machines can be used to do big things. And finally, studying inventions and technology will not only help us understand the technological world in which we live, but will help us realize that God created man to be creative just like Him.

It's fun to teach physics. It's interesting too. Energy and motion are all around us. We use technology and inventions every day. Finally, teaching physics is easy. You won't have to try to find strange materials for experiments or do dangerous things to learn about physics. Physics is as close as your child's toy box or the telephone—it's the rainbow in the sky and it's the light bulb in the lamp. So enjoy your study of the physical world.

How Do I Teach Science?

In order to teach any subject, you need to understand that people learn in different ways. Most people, and children in particular, have a dominant or preferred learning style in which they absorb and retain information more easily.

If a student's dominant style is:

Auditory
He needs not only to hear the information but he needs to hear himself say it. This child needs oral presentation as well as oral drill and repetition.
Visual
She needs things she can see. This child responds well to flashcards, pictures, charts, models, etc.
Kinesthetic
He needs active participation. This child remembers best through games, hands-on activities, experiments, and field trips.

Also, some people are more relational while others are more analytical. The relational student needs to know why this subject is important and how it will affect him personally. The analytical student, however, wants just the facts.

If you are trying to teach more than one student, you will probably have to deal with more than one learning style. Therefore, you need to present your lessons in several different ways so that each student can grasp and retain the information.

Grades 3–8

Each lesson should be completed by all upper elementary and junior high students. This is the main part of the lesson containing a reading section, a hands-on activity that reinforces the ideas in the reading section (blue box), and a review section that provides review questions and application questions (red box).

Grades 6–8

For middle school/junior high age students, we provide a "Challenge" section that contains more challenging material as well as additional activities and projects for older students (green box).

We have included periodic biographies to help your students appreciate the great men and women who have gone before us in the field of science.

We suggest a threefold approach to each lesson:

Introduce the topic

We give a brief description of the facts. Frequently you will want to add more information than the essentials given in this book. In addition to reading this section aloud, you may wish to do one or more of the following:

- Read a related book with your students.
- Write things down to help your visual students.
- Give some history of the subject. We provide some historical sketches to help you, but you may want to add more.
- Ask questions to get your students thinking about the subject.
- The "FUN FACT" section adds fun or interesting information.

Make observations and do experiments

- Hands-on projects are suggested for each lesson. This section of each lesson may require help from the teacher.
- Have your students perform the activity by themselves whenever possible.

Review

- The "What did we learn?" section has review questions.
- The "Taking it further" section encourages students to
  - Draw conclusions
  - Make applications of what was learned
  - Add extended information to what was covered in the lesson

By teaching all three parts of the lesson, you will be presenting the material in a way that all learning styles can both relate to and remember.

Also, this approach relates directly to the scientific method and will help your students think more scientifically. The *scientific method* is just a way to examine a subject logically and learn from it. Briefly, the steps of the scientific method are:

1. Learn about a topic.
2. Ask a question.
3. Make a hypothesis (a good guess).
4. Design an experiment to test your hypothesis.
5. Observe the experiment and collect data.
6. Draw conclusions. (Does the data support your hypothesis?)

Note: It's okay to have a "wrong hypothesis." That's how we learn. Be sure to help your students understand why they sometimes get a different result than expected.

Our lessons will help your students begin to approach problems in a logical, scientific way.

How Do I Teach Creation vs. Evolution?

We are constantly bombarded by evolutionary ideas about the earth in books, movies, museums, and even commercials. These raise many questions: Do physical processes support evolutionary theories? Do physical laws support an old earth? Do changes in the magnetic field support an old earth? The Bible answers these questions, and this book accepts the historical accuracy of the Bible as written. We believe this is the only way we can teach our children to trust that everything God says is true.

There are five common views of the origins of life and the age of the earth:

Historical biblical account	Progressive creation	Gap theory	Theistic evolution	Naturalistic evolution
Each day of creation in Genesis is a normal day of about 24 hours in length, in which God created everything that exists. The earth is only thousands of years old, as determined by the genealogies in the Bible.	The idea that God created various creatures to replace other creatures that died out over millions of years. Each of the days in Genesis represents a long period of time (day-age view) and the earth is billions of years old.	The idea that there was a long, long time between what happened in Genesis 1:1 and what happened in Genesis 1:2. During this time, the "fossil record" was supposed to have formed, and millions of years of earth history supposedly passed.	The idea that God used the process of evolution over millions of years (involving struggle and death) to bring about what we see today.	The view that there is no God and evolution of all life forms happened by purely naturalistic processes over billions of years.

Any theory that tries to combine the evolutionary time frame with creation presupposes that death entered the world before Adam sinned, which contradicts what God has said in His Word. The view that the earth (and its "fossil record") is hundreds of millions of years old damages the gospel message. God's completed creation was "very good" at the end of the sixth day (Genesis 1:31). Death entered this perfect paradise *after* Adam disobeyed God's command. It was the punishment for Adam's sin (Genesis 2:16–17; 3:19; Romans 5:12–19). Thorns appeared when God cursed the ground because of Adam's sin (Genesis 3:18).

The first animal death occurred when God killed at least one animal, shedding its blood, to make clothes for Adam and Eve (Genesis 3:21). If the earth's "fossil record" (filled with death, disease, and thorns) formed over millions of years before Adam appeared (and before he sinned), then death no longer would be the penalty for sin. Death, the "last enemy" (1 Corinthians 15:26), diseases (such as cancer), and thorns would instead be part of the original creation that God labeled "very good." No, it is clear that the "fossil record" formed sometime *after* Adam sinned—not many millions of years before. Most fossils were formed as a result of the worldwide Genesis Flood.

When viewed from a biblical perspective, the scientific evidence clearly supports a recent creation by God, and not naturalistic evolution and millions of years. The volume of evidence supporting the biblical creation account is substantial and cannot be adequately covered in this book. If you would like more information on this topic, please see the resource guide in the appendices To help get you started, just a few examples of evidence supporting biblical creation are given below:

Evolutionary Myth: Physical processes support evolution.

The Truth: Much of what scientists observe directly contradicts the ideas of evolution. Certain physical properties have been observed and tested to the point that they have been declared to be physical laws. The first law of thermodynamics states that matter and energy cannot be created or destroyed; they can only change form. There is no mechanism in nature for creating either energy or matter. Therefore, evolutionists cannot explain how all of the matter and energy in the universe came to be. This is a topic most evolutionists tend to ignore. The Bible tells us that God created it all and set it in motion.

The second law of thermodynamics states that all systems move toward a state of maximum entropy. This means that everything moves toward total disorganization and equilibrium. Heat moves from an area of higher temperature to an area of lower temperature, and organized systems become disorganized. For example, an organized system of cells that makes up a living creature quickly becomes disorganized when that creature dies. A house left to itself will eventually crumble into dust. Everything around us says that without intervention, chaos and disorganization result. Evolutionists, however, believe that by accident, simple molecules and simple organisms combined to form more complex molecules and organisms. This flies in the face of the second law of thermodynamics and everything that is observed to happen naturally. The changes required for the formation of the universe, the planet earth and life, all from disorder, run counter to the physical laws we see at work today. There is no known mechanism to harness the raw energy of the universe and generate the specified complexity we see all around us.[1]

A third physical property that contradicts evolution is the small amount of helium in the atmosphere. Helium is naturally generated by the radioactive decay of elements in the earth's crust. Because helium is so light, it quickly moves up through the rocks and into the atmosphere. Helium is entering the atmosphere at about 13 million atoms per square inch per second (67 grams/second). Some helium atoms are also escaping the atmosphere into space, but the amount of helium escaping into space is only about 1/40th the amount entering the atmosphere. So, the overall amount of helium in the atmosphere is continually increasing. If you assume that helium cannot enter the atmosphere any other way, which is a reasonable assumption, then the amount of helium in the atmosphere indicates that the earth could be no more than two million years old, which is much less than the billions of years needed for evolution. This is a maximum age—the actual age could be much less since this calculation assumes that the original atmosphere had no helium whatsoever. Also, helium could have been released at a much greater rate during the time after the Genesis Flood. Therefore, the amount of helium in the atmosphere indicates a much younger earth than evolutionists claim.[2]

[1] John D. Morris, *The Young Earth* (Colorado Springs: Creation Life Publishers, 1994), p. 43. See also www.answersingenesis.org/go/thermodynamics.

[2] Ibid., pp. 83–85.

Evolutionary Myth: Changes in the earth's magnetic field indicate an earth that is billions of years old.

The Truth: Most scientists agree on some fundamental facts concerning the earth's magnetic field. The earth is a giant electromagnet. The earth is surrounded by a magnetic field that is believed to be generated by current flowing through the interior of the earth. And there is evidence that the magnetic field of the earth has reversed several times. Also, nearly everyone agrees that the magnetic field is decreasing. The disagreement between evolutionists and creationists concerns how long it takes for the earth's magnetic field to change and what caused or causes the changes. Evolutionists believe that the magnetic field slowly decreases over time, reverses, and then slowly increases again. There are some serious problems with this idea. First, when the magnetic field is very low the earth would have no protection from very harmful radiation from the sun. This would be detrimental to life on earth. Second, at the current rate of decay, the magnetic field of the earth would lose half its energy about every 1,460 years. If the rate of decay is constant, the magnetic field would have been so strong only 20,000 years ago that it would have caused massive heating in the earth's crust and would have killed all life on earth. This supports the idea of an earth that is only about 6,000 years old, as taught in the Bible.

Creationists believe that the magnetic field reversals happened very quickly, and that the decay rate is fairly constant. One study of a lava flow indicated that reversals occurred in only 15 days. Thus, the reversals likely happened as a result of the Genesis Flood when the tectonic plates were moving and the earth's crust was in upheaval.[3]

[3] Ibid., pp. 74–83.

Despite the claims of many scientists, if you examine the evidence objectively, it is obvious that evolution and millions of years have not been proven. You can be confident that if you teach that what the Bible says is true, you won't go wrong. Instill in your student a confidence in the truth of the Bible in all areas. If scientific thought seems to contradict the Bible, realize that scientists often make mistakes, but God does not lie. At one time scientists believed that the earth was the center of the universe, that living things could spring from non-living things, and that blood-letting was good for the body. All of these were believed to be scientific facts but have since been disproved, but the Word of God remains true. If we use modern "science" to interpret the Bible, what will happen to our faith in God's Word when scientists change their theories yet again?

Integrating the Seven C's

Throughout the *God's Design® for Science* series you will see icons that represent the Seven C's of History. The Seven C's is a framework in which all of history, and the future to come, can be placed. As we go through our daily routines we may not understand how the details of life connect with the truth that we find in the Bible. This is also the case for students. When discussing the importance of the Bible you may find yourself telling students that the Bible is relevant in everyday activities. But how do we help the younger generation see that? The Seven C's are intended to help.

The Seven C's can be used to develop a biblical worldview in students, young or old. Much more than entertaining stories and religious teachings, the Bible has real connections to our everyday life. It may be hard, at first, to see how many connections there are, but with practice, the daily relevance of God's Word will come alive. Let's look at the Seven C's of History and how each can be connected to what the students are learning.

Creation

God perfectly created the heavens, the earth, and all that is in them in six normal-length days around 6,000 years ago.

This teaching is foundational to a biblical worldview and can be put into the context of any subject. In science, the amazing design that we see in nature—whether in the veins of a leaf or the complexity of your hand—is all the handiwork of God. Virtually all of the lessons in *God's Design for Science* can be related to God's creation of the heavens and earth.

Other contexts include:

Natural laws—any discussion of a law of nature naturally leads to God's creative power.

DNA and information—the information in every living thing was created by God's supreme intelligence.

Mathematics—the laws of mathematics reflect the order of the Creator.

Biological diversity—the distinct kinds of animals that we see were created during the Creation Week, not as products of evolution.

Art—the creativity of man is demonstrated through various art forms.

History—all time scales can be compared to the biblical time scale extending back about 6,000 years.

Ecology—God has called mankind to act as stewards over His creation.

Corruption

After God completed His perfect creation, Adam disobeyed God by eating the forbidden fruit. As a result, sin and death entered the world, and the world has been in decay since that time. This point is evident throughout the world that we live in. The struggle for survival in animals, the death of loved ones, and the violence all around us are all examples of the corrupting influence of sin.

Other contexts include:

Genetics—the mutations that lead to diseases, cancer, and variation within populations are the result of corruption.

Biological relationships—predators and parasites result from corruption.

History—wars and struggles between mankind, exemplified in the account of Cain and Abel, are a result of sin.

Catastrophe

God was grieved by the wickedness of mankind and judged this wickedness with a global Flood. The Flood covered the entire surface of the earth and killed all air-breathing creatures that were not aboard the Ark. The eight people and the animals aboard the Ark replenished the earth after God delivered them from the catastrophe.

The catastrophe described in the Bible would naturally leave behind much evidence. The studies of geology and of the biological diversity of animals on the planet are two of the most obvious applications of this event. Much of scientific understanding is based on how a scientist views the events of the Genesis Flood.

Other contexts include:

Biological diversity—all of the birds, mammals, and other air-breathing animals have populated the earth from the original kinds which left the Ark.

Geology—the layers of sedimentary rock seen in roadcuts, canyons, and other geologic features are testaments to the global Flood.

Geography—features like mountains, valleys, and plains were formed as the floodwaters receded.

Physics—rainbows are a perennial sign of God's faithfulness and His pledge to never flood the entire earth again.

Fossils—Most fossils are a result of the Flood rapidly burying plants and animals.

Plate tectonics—the rapid movement of the earth's plates likely accompanied the Flood.

Global warming/Ice Age—both of these items are likely a result of the activity of the Flood. The warming we are experiencing today has been present since the peak of the Ice Age (with variations over time).

Confusion

God commanded Noah and his descendants to spread across the earth. The refusal to obey this command and the building of the tower at Babel caused God to judge this sin. The common language of the people was confused and they spread across the globe as groups with a common language. All people are truly of "one blood" as descendants of Noah and, originally, Adam.

The confusion of the languages led people to scatter across the globe. As people settled in new areas, the traits they carried with them became concentrated in those populations. Traits like dark skin were beneficial in the tropics while other traits benefited populations in northern climates, and distinct people groups, not races, developed.

Other contexts include:

Genetics—the study of human DNA has shown that there is little difference in the genetic makeup of the so-called "races."

Languages—there are about seventy language groups from which all modern languages have developed.

Archaeology—the presence of common building structures, like pyramids, around the world confirms the biblical account.

Literature—recorded and oral records tell of similar events relating to the Flood and the dispersion at Babel.

Christ

God did not leave mankind without a way to be redeemed from its sinful state. The Law was given to Moses to show how far away man is from God's standard of perfection. Rather than the sacrifices, which only covered sins, people needed a Savior to take away their sin. This was accomplished when Jesus Christ came to earth to live a perfect life and, by that obedience, was able to be the sacrifice to satisfy God's wrath for all who believe.

The deity of Christ and the amazing plan that was set forth before the foundation of the earth is the core of Christian doctrine. The earthly life of Jesus was the fulfillment of many prophecies and confirms the truthfulness of the Bible. His miracles and presence in human form demonstrate that God is both intimately concerned with His creation and able to control it in an absolute way.

Other contexts include:

Psychology—popular secular psychology teaches of the inherent goodness of man, but Christ has lived the only perfect life. Mankind needs a Savior to redeem it from its unrighteousness.

Biology—Christ's virgin birth demonstrates God's sovereignty over nature.

Physics—turning the water into wine and the feeding of the five thousand demonstrate Christ's deity and His sovereignty over nature.

History—time is marked (in the western world) based on the birth of Christ despite current efforts to change the meaning.

Art—much art is based on the life of Christ and many of the masters are known for these depictions, whether on canvas or in music.

Cross

Because God is perfectly just and holy, He must punish sin. The sinless life of Jesus Christ was offered as a substitutionary sacrifice for all of those who will repent and put their faith in the Savior. After His death on the Cross, He defeated death by rising on the third day and is now seated at the right hand of God.

The events surrounding the crucifixion and resurrection have a most significant place in the life of Christians. Though there is no way to scientifically prove the resurrection, there is likewise no way to prove the stories of evolutionary history. These are matters of faith founded in the truth of God's Word and His character. The eyewitness testimony of over 500 people and the written Word of God provide the basis for our belief.

Other contexts include:

Biology—the biological details of the crucifixion can be studied alongside the anatomy of the human body.

History—the use of crucifixion as a method of punishment was short-lived in historical terms and not known at the time it was prophesied.

Art—the crucifixion and resurrection have inspired many wonderful works of art.

Consummation

God, in His great mercy, has promised that He will restore the earth to its original state—a world without death, suffering, war, and disease. The corruption introduced by Adam's sin will be removed. Those who have repented and put their trust in the completed work of Christ on the Cross will experience life in this new heaven and earth. We will be able to enjoy and worship God forever in a perfect place.

This future event is a little more difficult to connect with academic subjects. However, the hope of a life in God's presence and in the absence of sin can be inserted in discussions of human conflict, disease, suffering, and sin in general.

Other contexts include:

History—in discussions of war or human conflict the coming age offers hope.

Biology—the violent struggle for life seen in the predator-prey relationships will no longer taint the earth.

Medicine—while we struggle to find cures for diseases and alleviate the suffering of those enduring the effects of the Curse, we ultimately place our hope in the healing that will come in the eternal state.

The preceding examples are given to provide ideas for integrating the Seven C's of History into a broad range of curriculum activities. We would recommend that you give your students, and yourself, a better understanding of the Seven C's framework by using AiG's *Answers for Kids* curriculum. The first seven lessons of this curriculum cover the Seven C's and will establish a solid understanding of the true history, and future, of the universe. Full lesson plans, activities, and student resources are provided in the curriculum set.

We also offer bookmarks displaying the Seven C's and a wall chart. These can be used as visual cues for the students to help them recall the information and integrate new learning into its proper place in a biblical worldview.

Even if you use other curricula, you can still incorporate the Seven C's teaching into those. Using this approach will help students make firm connections between biblical events and every aspect of the world around them, and they will begin to develop a truly biblical worldview and not just add pieces of the Bible to what they learn in "the real world."

UNIT 1

COMMUNICATIONS

LESSON 1

PRINTING PRESS

COMMUNICATIONS BREAKTHROUGH

SUPPLY LIST

Potatoes   Knife   Paper   Tempera paints   Rubber stamps   Ink pad

WHAT DID WE LEARN?

- What different forms of communication are commonly used? **Speaking /oral and written are the most common. Body language is also important. Smoke signals and drums are more unusual.**
- What was the very first human communication we know of? **God talking with Adam.**
- What are the three necessary parts of a printing press? **The type/letters, ink, and the press.**

TAKING IT FURTHER

- Why is the printing press such an important invention? **It allowed ideas and knowledge to be passed quickly and economically from one place or person to another.**
- How are modern presses different from the original printing presses? **They are much faster, computer controlled, and more accurate. There are many different technologies, some of which do not involve pressing ink at all. Many computer printers work by spraying jets of ink onto the paper. Most newspapers, magazines, and books are still printed on presses. However, the type is no longer set by hand. Instead photographic plates are generated by computers and ink is applied to the plates and then pressed onto the paper.**

TELEGRAPH

COMMUNICATION WITH WIRES

SUPPLY LIST

Copy of "Morse Code Puzzles"

MORSE CODE PUZZLES

1. **Have a happy day.**
2. **God loves you.**
3. **What hath God wrought?**
4. **There are 7 days in 1 week.**

What did we learn?

- What is the purpose of a telegraph system? **To send messages long distances in a short period of time using electricity.**
- What is the function of the telegraph key? **When pressed, it completes an electrical circuit, sending a pulse of electricity along a wire. This sends the message.**
- What is the function of the receiver? **The receiver produces an audible click with each pulse of electricity. This allows the person at the receiving end to hear the coded message.**
- What is Morse code? **A series of dots and dashes that represent each letter of the alphabet as well as numbers and several symbols.**
- What is the difference between a dot and a dash in Morse code? **A dot is a short electrical pulse and a dash is a longer electrical pulse.**
- Who is credited with inventing the telegraph? **Samuel F. B. Morse.**

Taking it further

- Why was the telegraph better than other methods of communication at that time? **It was much faster than any other method for delivering information.**
- What changes do you think occurred in society because of the invention of the telegraph? **There were many changes. One very important change was that news from all parts of the world was now available nearly instantaneously. Although the common person did not usually send or receive telegrams, reporters sent telegrams so people across the country and around the world would know when an important event took place.**

Telephone

Hello

Supply list

2 paper or Styrofoam cups Paper clips String
Supplies for Challenge: Clear plastic tubing (¼" diameter) Black paint
Desk lamp or flashlight Modeling clay Plastic wrap Cardboard box Tape

A Simple Telephone

- Explain how this homemade phone works. How is it similar to a real telephone? **The vibrations of your voice are captured in the cup and turned into vibrations along the string. These vibrations travel to the other cup, where they are turned back into sound waves. This is a very similar process to how a phone works, except that in your homemade phone the vibrations are turned into physical vibrations in the string instead of electrical impulses in a wire.**

What did we learn?

- What are the major functions of a telephone? **To translate sound into electrical signals and electrical signals back into sound.**
- What is the function of the keypad on a telephone? **To send routing information so the call goes to the correct telephone.**

- Who is credited with inventing the telephone? **Alexander Graham Bell.**
- What is a cell phone? **A phone that uses wireless radio signals to transmit information.**

Taking it further

- How did Alexander Graham Bell's work with deaf children influence his invention of the telephone? **His work gave him opportunity to study the workings of the human voice and a desire to replicate it.**
- How does a cellular phone operate differently from a regular telephone? **They do not have any wires. The voice sounds are translated into radio signals inside the phone then sent to satellites or transmission stations that send the radio signal to another cell phone or to a station that translates the radio signal into an electrical signal for a regular phone.**

Radio

No wires

Supply list

Portable radio   Metal file   Electrical wire   Large battery (4.5 to 6 Volts)   Tape
Supplies for Challenge: String   Modeling clay (two colors)   Copy of "Radio Tuner" worsheet

What did we learn?

- Who is the person most responsible for developing radio? **Guglielmo Marconi.**
- What are the main parts of a radio system? **Microphone/recording system, transmitter, tuner, and receiver.**
- What is the function of each of these parts? **The microphone system converts sound into electrical signals; the transmitter changes electrical signals into radio waves; the tuner filters out unwanted radio signals; the receiver converts radio waves into electrical signals and then into sound.**

Taking it further

- Why does a digital signal give you a better sound than an analog signal? **The analog signal can pick up background noise that distorts the sound. Digital signals are either there or not there—you don't get distortion.**
- What is an advantage of using a radio to send a message over using a traditional telephone? **The radio signal can go anywhere in a straight line from the transmitter. The traditional telephone signal can only go where wires have been installed.**
- Under what conditions might you choose to use a radio to contact someone? **If that person is moving or in a location where there are no telephone wires, although cellular phones are making traditional radio communication less necessary.**
- What is the main thing radio is used for today? **News and entertainment.**

Challenge: Radio Tuner worksheet

- Which string caused the tuner string to begin to swing? **Only the 15 inch string should cause the tuner string to swing.**

- How does this experiment demonstrate how a radio tuner works? **Each string has a natural frequency at which it vibrates. The tuner will begin to vibrate when it is close to something that is vibrating at its natural frequency. This is called resonance and is how the tuner works in your radio.**

Television

Pictures in your home

Supply list

Red, green, and blue cellophane or plastic wrap   Flashlight

Supplies for Challenge: Television remote control (if available)

What did we learn?

- How is a television different from a radio? **It produces visual images in addition to sound.**
- What is the picture tube in a traditional TV properly called? **Cathode ray tube/CRT.**
- What kind of information is transmitted by TV stations? **Sound along with red, green, and blue image information are all transmitted as a single, combined signal.**
- How does your TV produce a visual image on the screen? **The red, green, and blue signals are used to control electron emitting guns in the back of the CRT. These streams of electrons cause dots of phosphor to glow on the screen, thus producing the image.**

Taking it further

- How do you think your TV changes stations? **Just like in a radio, a TV has a tuner that only works at particular frequencies.**
- Why do you see a complete image when there are really just thousands of little red, green, and blue dots on the screen? **The dots are so small and so close together that your brain blends them together into one image.**
- What are some advantages that HDTV has over traditional TV? **HDTV provides a signal with better picture and sound quality.**

Fax Machine

Sending a copy

Supply list

2 copies of "Fax Machine Template"

Supplies for Challenge: 2 or more copies of "Fax Machine Template"   Magnifying glass   Ruler

Copy from a fax machine (if available)

What did we learn?

- What does fax mean? **It is short for facsimile, which means an exact copy.**
- What is the purpose of a fax machine? **To send a copy of an image that can be printed out.**
- How does a fax machine send a copy of an image? **It breaks the image down into millions of tiny squares and translates each square into an electrical signal for black or white and sends that signal to another fax machine over the telephone lines, where the receiving machine converts the electrical signals back into an image.**

Taking it further

- Why might a business use a fax machine? **To send or receive documents or pictures more quickly than through the mail.**
- How is a fax machine similar to a telephone? **It sends information as electrical signals over the same lines as a telephone. It has transmission and receiver hardware.**
- How is a fax machine different from a telephone? **A fax machine converts picture/image data into electrical signals, whereas a telephone converts sound data into electrical signals.**
- What technology is replacing some of the job of a fax machine? **The Internet and email.**

Challenge: Increasing Resolution

- **Answers will vary, but you probably need at least 256 squares to even come close to a circle. But a 32 by 32 grid, or 1024 squares, will be much better.**

Computer

The ultimate in communications?

Supply list

Paper and pencil

Supplies for Challenge: Copy of "Computer Architecture" worksheet

Computers in Action

- Your list could be very long if you tried to list everything, but try to list at least 10–15 ways that computers are used around you. **Some ideas could include checking out at the store, paying at the pump when you buy gasoline; computers control airplanes and space ships, they are used to track shipments, they keep track of inventory in stores, they are used for banking, and computers are used to design video games. Your car probably has a computer that controls the functions of the engine. I am sure you can think of many more ideas.**

What did we learn?

- What is the "brain" of a computer? **The CPU—central processing unit.**
- What is RAM? **Random Access Memory.**
- What is RAM used for? **To store commands and information that have been recently used, or will soon be used, by the CPU.**
- What were computers designed for initially? **To perform mathematical calculations.**

- What are some other functions of computers today? **Communication, word processing, controlling of many items, entertainment, etc.**

Taking it further

- How are binary codes, 1s and 0s, similar to Morse code? **The 1s and 0s represent information that is passed from one part of the computer to another, just as in Morse code where dots and dashes represent information that is passed from one location to another.**
- What new application can you think of for a computer? **Answers will vary.**
- What would you use a computer for if you could? **Answers will vary.**

Challenge: Computer Architecture worksheet

Monitor: **Output** Keyboard: **Input** Printer: **Output** Mouse: **Input**
Modem: **Input & output** Microphone: **Input** Joystick: **Input** RAM: **Input & output**
CD/DVD: **Input (& output if it is a CD/DVD writer)** Hard drive: **Input & output**
Network card: **Input & output** Flash drive: **Input & output**

Quiz 1 Communications

Lessons 1–7

Mark each statement as either True or False.

1. _**T**_ The printing press may be the most important invention in the last 1000 years.
2. _**F**_ The type for modern printing presses is set by hand.
3. _**F**_ The telegraph sends voice messages.
4. _**T**_ Most communication devices have a transmitter and a receiver.
5. _**T**_ Morse code is a series of long and short signals.
6. _**F**_ Samuel Morse invented the telephone.
7. _**T**_ A cellular telephone converts sound to radio waves.
8. _**T**_ Digital radio signals are usually clearer than analog signals.
9. _**T**_ Radio signals can reach outer space.
10. _**F**_ Red, white and blue are the three colors used on a TV.
11. _**T**_ The CRT is the picture tube in a television.
12. _**F**_ Tuners are needed for radio but not for television.
13. _**F**_ A fax machine's primary purpose is to send sound information.
14. _**T**_ A fax changes pictures into electrical signals.
15. _**F**_ A fax machine is just a telephone.
16. _**T**_ The CPU is the brain of a computer.
17. _**F**_ The RAM is the remote adding memory of a computer.
18. _**T**_ Computers are ideal for mathematical calculations.
19. _**F**_ Communication has changed slowly over the past 100 years.
20. _**T**_ Morse, Bell, Gutenberg, and Marconi are all important inventors.

Challenge Questions

Mark each statement as either True or False.

21. _T_ Different fonts give the printed word a different feel.
22. _F_ Fiber optic cables are used to transmit electrical signals.
23. _T_ Fiber optic signals are faster and more accurate than signals carried on wires.
24. _T_ A radio tuner uses materials that vibrate at certain frequencies.
25. _F_ A television remote control is always very complicated.
26. _F_ Curved lines transmit better on a FAX than straight lines.
27. _T_ The resolution of a FAX helps determine the quality of the copy.
28. _F_ A computer monitor is usually an input device.
29. _T_ A computer printer is an output device.
30. _T_ A computer mouse is an input device.

UNIT 2
TRANSPORTATION

STEAM ENGINE

A NEW AGE OF POWER

SUPPLY LIST

Aluminum foil String Scissors Tape Oven mitts

Supplies for Challenge: Research materials on the Industrial Revolution and/or the steam engine

WHAT DID WE LEARN?

- How does steam make things move? **As water changes to steam it expands, creating pressure. This pressure makes things move.**
- Who first invented the steam engine? **James Watt.**
- How is steam still used in industry today? **One use is to drive turbines in power stations.**
- How did Robert Fulton use the steam engine? **To drive the paddle wheel of a boat.**

TAKING IT FURTHER

- How is a steam engine different from a steam pump? **The pump uses steam to move a piston up and down and this up-and-down movement moves a pump up and down. An engine converts the up-and-down movement into rotational motion so it can be used to drive anything.**
- Why is the steam engine credited with sparking the Industrial Revolution? **The steam engine provided the power needed to drive the machinery that was used in all the industries, making large factories and mass production possible.**

TRAIN

FASTER THAN A HORSE AND CART

SUPPLY LIST

Soda straws Tape Tagboard or poster board Toy train Craft sticks

Supplies for Challenge: 4 or more block magnets 2 nickels Tape Tagboard or poster board

WHAT DID WE LEARN?

- What invention helped spur the invention of the train? **The steam engine.**
- How long did steam engines dominate the train industry? **For over 100 years.**

- What kinds of trains are common today? **Diesel and electric trains.**
- Why are electric trains preferred for passenger travel? **They are quieter and cleaner than diesel trains.**

Taking it further

- Why are diesel trains used more often than electric trains for transporting freight? **It is not practical to run electric lines in many areas or over long distances.**
- What weather conditions must be considered when building railroad bridges? **Wind can be a major problem for bridges. Depending on the location, snow might need to be considered. Also, near the ocean, waves and hurricanes may be considerations.**
- How is a truss bridge similar to an arch bridge? **The arch design and the triangular truss both distribute the weight on the bridge in such a way as to actually strengthen the design rather than weaken it.**

Internal Combustion Engine

Gasoline and Diesel

Supply list

Copy of "Internal Combustion Engine Pattern" Poster board or tagboard
Supplies for Challenge: 6 metal bolts Measuring cup Sauce pan 3 cereal bowls
Copy of "Cooling an Engine" worksheet

What did we learn?

- What is an internal combustion engine? **An engine that uses the force of an internal explosion of fuel to produce movement.**
- What are the main fuels used in combustion engines today? **Gasoline and diesel fuel.**
- What are the four stages or functions of the four stroke engine? **Intake, compression, power, and exhaust.**

Taking it further

- How does the movement of a piston turn a car's wheels? **The pistons are connected to a shaft that is part of the drive train. Through a series of gears, the up-and-down movement of the pistons is converted into circular movement in the wheels.**
- How is the action of an engine similar to the action of riding a bicycle? **The movement of the piston is similar to the up-and-down movement of the rider's legs.**

Challenge: Cooling an Engine worksheet

- Which bolts were the coolest/warmest? **The water-cooled bolts will be coolest; the bolts just sitting in the air will be warmest.**
- How was the heat removed from each set of bolts? **The first bolts transferred heat to the water, second bolts transferred heat to the air as it moved past, third bolts transferred heat to air around them.**
- How is this experiment similar to what occurs inside an engine? **The first bolts are similar to a water-cooled car engine, second bolts similar to air-cooled lawn mower engine, third bolts similar to a lawn mower engine that is idling or not moving.**

Automobile

Dad, can I borrow the car?

Supply list

Paper Markers

Supplies for Challenge: Research materials on hybrid cars

What did we learn?

- Why were internal combustion engines better than steam engines for automobiles? **Steam engines took too long to heat up and required too much fuel to be carried along.**
- What are some alternative fuel sources to gasoline and diesel fuel? **Electricity, biodiesel, ethanol, and hydrogen.**
- What are some parts of the automobile that have been improved over the years? **Transmission, steering, suspension, ignition, and safety.**
- What was Henry Ford's major contribution to the automobile industry? **The invention of the assembly line.**

Taking it further

- What kinds of vehicles use internal combustion engines other than cars and trucks? **Bus, motorcycle, fire engine, construction equipment, recreational vehicle, farm equipment, etc.**
- What factors must be considered when designing an automobile? **Function, fuel economy, safety, comfort, and cost are some of the factors.**
- What other industries have benefited from the automobile industry? **Nearly every industry uses assembly-line techniques that were started by Henry Ford.**

Jet Engine

Moving air

Supply list

Copy of "Jet Engine" worksheet

Supplies for Challenge: Copy of "Boyle's Law" worksheet

Jet Engine worksheet

- **See student manual, p. 54, for labeled jet engine.**
- Combustion chamber: **Here the superheated air is mixed with some kind of fuel. This fuel is usually jet fuel, but can also be kerosene, propane, or other fuel. When the fuel is mixed with the superheated air, it burns very quickly.**
- Fan: **Sits in front of the compressor to help draw in the air.**
- Compressor: **Draws in air and compresses it up to 40 times, thus superheating the air.**

- Turbine: **A series of blades that turn when the exhaust gases push against them.**
- Fuel injectors: **Spray fuel in to the combustion chamber to mix with air and then burn.**
- Bypass chamber: **Where much of the air goes that is forced in by the fan; the air goes around the compressor and out the back.**
- Describe why a turbo-fan jet engine provides more thrust than an ordinary turbo jet engine. **A turbo-fan engine has a large fan in front of the compressor to help draw in the air. The fan draws in much more air than the compressor can handle. Much of the air is forced around the compressor and out the back, like a simple jet engine, so two sources are pushing air/gases out the back instead of only one.**

What did we learn?

- How does a simple jet engine work? **Air is drawn in and compressed, then forced out the back of a hollow tube at a higher speed.**
- What are the three main parts of a turbo jet engine? **The compressor, the combustion chamber, the turbine.**
- What is the purpose of jet fuel in a turbo jet engine? **The burning of the fuel causes the gases to expand, move the turbine, and exit the engine at great speeds, creating thrust.**
- In what applications are jet engines used, other than for airplanes? **Helicopters, tanks, and power plants.**

Taking it further

- Why were steam engines abandoned as possible airplane engines? **They were too heavy for the amount of thrust they provided.**
- Why are jet engines not used in most cars and trucks? **They are too expensive and provide more thrust than is necessary.**
- How is a jet engine power plant similar to a coal power plant? How are they different? **They both create electricity by the turning of turbines. The way in which they turn the turbine is different. The coal power plant uses coal to heat water into steam and the steam turns the turbine. The jet engine uses fuel to heat gases and the expanding gases turn the turbine.**
- How does a turbo-fan engine create more thrust than the straight turbo jet engine? **The fan draws in more air than can be sent through the compressor. This extra air is compressed as it flows around the engine and exits at a higher speed than it entered, thus increasing the overall thrust of the engine.**

Challenge: Boyle's Law worksheet

- Explain how Boyle's law is applied to the compression of air in a turbo jet engine: **Boyle's law—The compression of the air results in greatly increased temperature which ignites the fuel that is injected into the combustion chamber.**
- Explain how Newton's third law of motion applies to the turbo jet engine: **Newton's third law—The force of the air exiting from the back of the engine produces an equal force on the airplane in the forward direction, thrusting it forward.**
- How do these two laws work together to increase the thrust of the turbo jet engine? **The burning fuel expands rapidly, creating thrust as it exits the back of the engine.**

LESSON 13

Airplane

I can fly!

Supply list

Copy of "Airplane Movements" worksheet

Airplane Movements worksheet

- Identify whether each picture below demonstrates pitch, roll, or yaw, then fill in the blanks below.
  A. **Yaw** B. **Pitch** C. **Roll**

1. If the pilot pushes the nose of the plane downward, it will experience a change in what? **Pitch.**
2. If the plane dips the left wing, it will experience what kind of movement? **Roll.**
3. Which kind of movement will be needed to increase the plane's altitude? **An increase in pitch.**
4. Which kind of movement will be needed to make a course correction because of prevailing winds? **A change in yaw.**

What did we learn?

- Explain each of the four forces that affect airplane flight. **Lift is the pressure generated by air flowing over the airfoil-shaped wing. Weight is the resistance to lift that occurs because of gravity's pull on the mass of the plane. Thrust is the forward force generated by the engine. Drag is the stopping force of air resistance.**
- Explain how Bernoulli's principle causes lift. **The air flowing over the top of the wing flows faster than the air under the wing. This reduces the air pressure above the wing with respect to the air pressure below the wing; thus the air under the wing pushes up on the wing causing lift.**
- Explain the three different ways that a plane moves in the air. **Pitch is moving up and down, roll is tipping side to side, and yaw is turning right or left.**

Taking it further

- How is Newton's third law of motion applied to airplanes? **The thrust of the engines is generated by air exiting the back of the engine, thus creating a forward reaction.**
- What will happen to an airplane if the power is reduced in the engine? **The thrust will decrease while the drag remains the same so the plane will slow down. If the plane slows down enough, the lift may also decrease and the plane will lose altitude.**

Rocket Engine

Reaching for space

Supply list

Balloons Tagboard Tape Soda straw String
Supplies for Challenge: Research materials on the V-2 rocket

What did we learn?

- Who first explained the scientific principles behind rockets? **Sir Isaac Newton.**
- What is the third law of motion? **For every action there is an equal and opposite reaction.**
- Who is considered the father of modern rocketry? **Robert Goddard.**
- What is a propellant? **The fuel and oxidizer used in a rocket.**
- What are the three types of rocket fuel used today? **Solid, cryogenic, and hypergolic.**

Taking it further

- What are two important things that an engineer must consider when designing a rocket engine? **The thrust needed to lift the total weight and the type of fuel needed.**
- Why do you think model rocket engines are made from solid rocket fuel? **Solid fuel is easy to store and easy to ignite with an electrical charge; it is inexpensive; model rockets don't need to be directed and do not go into space.**
- Many rockets burn their fuel in two or three stages. Why might they be designed this way? **On the launch pad, the engine must have enough thrust to lift the total weight of all fuel as well as the rocket, people and other materials on board, so a large engine is needed. But once much of the fuel is used up, a smaller engine can be used, so the first stage is ejected and the smaller engine takes over. This allows for less fuel consumption overall.**
- How can a rocket engine work in space where there is no air? **The oxygen for burning the fuel is in the rocket engine so no outside air is needed for combustion. Air is not needed for the Third Law of Motion to work. The gases are still exiting the back of the engine, which still propels the rocket forward even when there is no atmosphere.**
- Would you expect a rocket engine to be more or less efficient in space? Why? **It is more efficient in space because there is no air resistance.**

Spacecraft

Escaping from earth

Supply list

Drawing materials

Supplies for Challenge: Research materials on the International Space Station
Copy of "International Space Station Facts" worksheet

What did we learn?

- What is a spacecraft? **Any object that was designed to be used beyond earth's atmosphere.**
- List three different kinds of spacecraft? **Satellites, probes, and a manned craft.**
- What two countries have led the way in the development of spacecraft? **Soviet Union/Russia and the United States.**
- What scientist launched the first rocket into space? **Werner Von Braun.**
- What is the Cassini-Huygens probe? **A space probe sent to study Saturn and its moons.**

Taking it further

- Why does the Huygens space probe need a heat shield? **It was sent to the surface of Titan, Saturn's largest moon. Titan has an atmosphere, so a heat shield was needed to protect the probe from the heat generated by friction during entry through the atmosphere.**
- Why does Cassini use a nuclear power source instead of solar panels? **It is being used too far from the sun for solar panels to be effective.**
- What is one advantage to a space telescope such as Hubble? **There is no distortion or interference from the earth's atmosphere in space.**

Challenge: International Space Station Facts worksheet

1. **404,069 pounds; may change as more parts are added.**
2. **Solar arrays/solar energy.**
3. **15,000 cubic feet.**
4. **Nov. 2, 2000.**
5. **Normal clothing.**
6. **EMU—extravehicular mobility unit space suit.**
7. **2.5 hours per day.**
8. **Email—updated 3 times per day.**
9. **NASA—National Aeronautics and Space Administration.**
10. **Oct. 1, 1958.**
11. **Peggy Whitson—biochemist.**
12. **Answers will vary.**

Hovercraft

Floating on air

Supply list

Styrofoam or paper plate   Soda straw

Supplies for Challenge: Modeling clay   Cardboard   Toothpicks   Pitcher

What did we learn?

- What is an ACV? **Air-cushioned vehicle or hovercraft.**
- How are hovercraft different from other vehicles? **They float on a cushion of air so they can travel on both land and water.**
- Why are hovercraft preferable to traditional boats in certain areas? **They can travel in water that is too shallow for traditional boats. Also, they can go up onto the land for loading and unloading.**
- Why are hovercraft more energy efficient than many other vehicles? **They greatly reduce the friction that other vehicles experience.**

Taking it further

- Which of Newton's laws of motion are most easily seen in the operation of hovercraft? **Newton's third law—for every action there is an equal and opposite reaction.**
- How are harrier jets similar to hovercraft? **They both use downward moving air to create lift.**

Challenge: Reducing Friction

- **Circle and tear drop allow water to flow smoothly. Square and star hinder flow. Vehicles have smooth rounded surfaces to aid with air or water flow.**

Transportation

Lessons 8–16

List the engine type(s) that each statement below applies to.

1. **_S, J_** Turns a turbine
2. **_IC, J, R, S—wood and coal cannot burn without oxygen_** Mixes fuel with oxygen
3. **_J, R_** Hot gases exiting the engine push the vehicle forward
4. **_S can be turbine or piston, IC_** Expanding gas pushes pistons up and down
5. **_S_** Requires wood or coal to heat water
6. **_S, IC, J, R_** Uses Newton's Third Law of Motion
7. **_IC_** May require spark plugs
8. **_R_** Hydrogen is the primary fuel source **(hydrogen combustion engines are being designed for some cars)**
9. **_R_** Provides the most thrust
10. **_S_** Oldest engine design
11. **_S_** Spurred the invention of the train
12. **_IC_** Commonly used for trains today

Match the inventor with his invention.

13. **_B_** James Watt
14. **_D_** Robert Fulton
15. **_E_** Robert Goddard
16. **_C_** Rudolph Diesel
17. **_F_** Gottlieb Daimler
18. **_H_** Henry Ford
19. **_G_** Orville and Wilbur Wright
20. **_A_** Frank Whittle

Challenge questions

Short answer:

21. Explain how magnetic forces are used to move a maglev train. **Electromagnets in the track create opposite magnetic fields to the magnets in the trains. This creates repelling forces that push the train forward.**
22. Explain how the steam engine affected the industrial revolution. **Many industries were able to use the steam engine to run mills or other equipment thus spurring on industrial development. The steam engine was quickly adapted to trains.**
23. Give two ways that heat is reduced or eliminated from internal combustion engines. **Lubricants help reduce friction; radiators and water circulation removes heat.**
24. Explain how increasing the weight of an airplane affects the size of the engines needed. **Increased weight requires that the engines be able to provide more thrust because more thrust is needed to provide more lift to counter the additional weight.**
25. How does Bernoulli's principle apply to hydrofoils? **Water flowing over the top of the waterfoils moves faster than water under the waterfoils. This causes lift, allowing the boat to move above the surface of the water.**

Unit 3
Military Inventions

Historical Military Weapons

The physics of war

Supply list

Small block Ruler Small eraser

What did we learn?

- What were some of the earliest weapons invented? **Bows and arrows, shields, swords, spears, and knives.**
- What military invention was the most revolutionary? **Gunpowder.**
- What kinds of weapons were used by the Romans? **Catapults, ballista, swords, and spears.**

Taking it further

- What are some ways that military technology has helped people? **Guns and knives are useful for hunting and protection. Rockets are used for launching satellites and for scientific exploration. Sonar and radar have many civilian applications as well, such as navigation and sea exploration.**
- Name at least two inventions not listed in the lesson that have both military and civilian applications. **Airplanes, communication equipment, satellite technology, and space technology are just a few examples.**

Gunpowder

It's explosive

Supply list

Unpopped popcorn kernels Rice Flour

What did we learn?

- Who first discovered gunpowder? **The Chinese.**
- How was gunpowder first used? **For fireworks.**
- What is ballistics? **The science that deals with the motion of projectiles.**
- What were some problems with the matchlock gun? **You had to light it ahead of time, sometimes the fuse went out, the glow could be seen by the enemy at night.**

Taking it further

- How was gunpowder made more reliable? **By mixing the ingredients in a liquid and letting the mixture dry into pellets.**
- How is a cannon similar to a catapult? **Both shoot projectiles a long distance.**
- How is a cannon different from a catapult? **It uses the explosion of ignited gunpowder to force the cannonball to fly instead of using tension.**
- How might gunpowder be used for peaceful purposes? **Fireworks, hunting, and explosives for mining or earthmoving.**

Challenge: Improving Accuracy

- **If there are any imperfections in the bullet, it will not fly straight. Similarly, if there are any imperfections in the gun barrel the bullet will not fly straight. Spinning the bullet offsets some of these imperfections, thus increasing accuracy.**
- **Putting a spin on a football helps it fly straighter and improves accuracy.**

Tank

A modern day chariot

Supply list

2 spools of thread　Modeling clay　Craft sticks　Heavy book

The Advantage of Tracks

- How are the indentations different? **The indentations should not be as deep and should be more spread out.**
- Why are the indentations shallower when the sticks are used? **The weight from the book is not concentrated in four small areas, but is spread out.**
- How is this similar to the treads on a tank? **The weight of the tank is spread out along the tracks so there is no single area that digs into the ground, giving it better mobility.**

What did we learn?

- What is the purpose of a tank in warfare? **To provide a large amount of firepower, to move soldiers in a protected environment, and to give the army mobility.**
- What are two of the earliest war vehicles? **Chariots and war elephants.**
- What invention made traditional armor useless? **Firearms/gunpowder.**
- What invention made tanks possible? **The Caterpillar track system.**
- When were the first military tanks invented? **During World War I, 1915–1916.**

Taking it further

- Why is the track system necessary for a tank? **The track spreads out the weight so that a very heavy vehicle will not get stuck as easily as a wheeled vehicle would.**
- How can airplanes take the place of tanks in some battles? **They can get in and out faster than tanks.**
- Why are tanks still necessary? **They can clear a path for infantry. Airplanes cannot do this as well.**

Challenge: Gasoline vs. Diesel

- **Answers will vary, but should reflect thought concerning which advantages outweigh the disadvantages.**

Lesson 20 Submarine

Underwater boat

Supply list

2-liter plastic bottle with cap   Ketchup packets

What did we learn?

- What is a submarine? **An underwater vehicle.**
- What were some of the problems that had to be solved in order to make a successful submarine? **Buoyancy, steering under water, water-tight hull under great pressure, fresh air, power underwater, communications under water.**
- What is ballast? **Extra material that is carried in the vehicle that can be eliminated to make the vehicle lighter.**
- What material is used as ballast on a submarine? **Seawater.**
- What kind of vehicle used ballast before submarines? **Hot air balloons.**
- When was the first submarine used in a battle? **In the American Revolutionary War—1776.**

Taking it further

- How were the first submarines powered? **Human power—foot or hand cranks were used to turn propellers.**
- List three advantages of nuclear submarines. **They are quieter, can stay submerged longer, have more living space, and no diesel fumes.**
- Why does a nuclear submarine have more living space than a diesel submarine? **The nuclear reactor is heavier than a diesel engine; therefore, the submarine must be bigger to reduce its density.**
- How does ballasting work on a submarine? **Ballast tanks can be filled with either water or air. When they are filled with air, the sub becomes lighter than water. When they are filled with water, the sub becomes heavier than water. The submarine carries compressed air that can be released into the ballast tanks when the sub needs to surface.**
- Why are submarines useful in warfare? **They are difficult to detect and can sneak up on the enemy.**
- Why are electric batteries used under water rather than diesel engines? **Primarily because they are quieter and harder to detect.**

LESSON 21

Radar & Sonar

Reflectors

Supply list

Copy of "Sonar" worksheet

Sonar worksheet

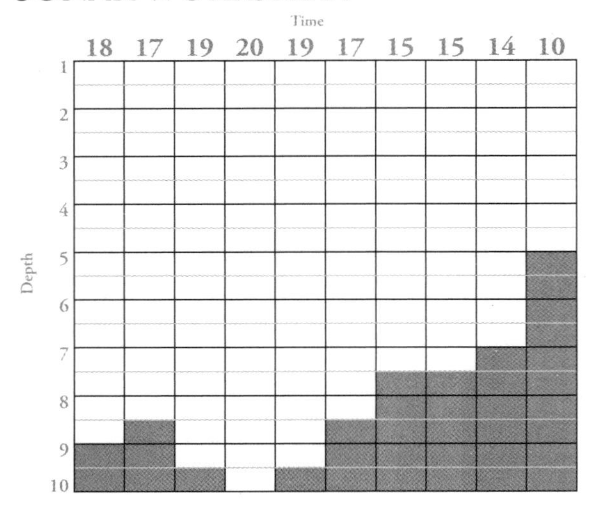

What did we learn?

- What is radar? **Radio detection and ranging—it is the use of radio waves to detect objects.**
- What is sonar? **Sound navigation and ranging—it is the use of sound waves to detect objects.**
- When was radar first developed? **From 1935–1940.**
- What are the two main parts of a radar system? **The transmitter and the receiver.**

Taking it further

- Why do military submarines have a passive radar system? **To listen for sounds generated by other submarines without giving their own position away.**
- Why must a radar system divide by two the time from the transmission of the signal until the echo is received in order to determine the distance to the object? **Because the sound has traveled from the radar system to the object and back, so it has traveled twice the distance that the object is from the radar system.**
- How can a sonar system be used to detect objects without being confused by the floor of the ocean? **The system can be programmed to look only at echoes that come from objects that are to the sides or above it. Also, if the floor of the ocean has been mapped, the computer can be programmed to filter out the ocean floor echoes.**
- Why might it be important for a commercial airline to use radar to detect its exact altitude? **If the pilot is landing in poor visibility, he/she needs to know exactly where the ground is even if he/she cannot see it.**

Challenge: Doppler Radar

- **Police speed gun, measuring speed in athletic events, and measuring movements in space.**

QUIZ 3

Military Inventions

Lessons 17–21

Choose the best answer for each question.

1. _A_ Which invention most revolutionized military weapons?
2. _C_ Which is not an ingredient in gunpowder?
3. _B_ Who first developed gunpowder?
4. _D_ What branch of physics studies the movement of projectiles?
5. _A_ What part of a tank makes it unique—different from other vehicles?
6. _A_ What vehicle was designed for underwater warfare?
7. _B_ What physical property allows submarines to rise and sink in the water?
8. _D_ What substance is used as ballast on a submarine?
9. _B_ What invention allows for the detection of objects using radio waves?
10. _C_ What invention allows for the detection of objects using sound waves?
11. _C_ Which kind of radar system listens for signals only?
12. _A_ Which invention required submarines to be made larger?
13. _D_ Which military invention was first used in World War I?
14. _A_ Which of the following served some of the same military purposes as a tank?

Challenge Questions

Choose the best answer for each statement or question.

15. _C_ Which type of lever is best for a catapult?
16. _C_ A rifled gun barrel shoots more accurately because it causes the bullet to _____.
17. _B_ A diesel engine might be better in a tank than a gasoline engine because it is _____.
18. _A_ At what depth is a submarine designed to operate?
19. _D_ The Doppler effects cause which characteristic of an energy wave to change?
20. _A_ Doppler radar is used to track the movement of _____.

UNIT 4
MODERN CONVENIENCES

ELECTRIC LIGHT

A BRIGHT IDEA

SUPPLY LIST

Whatever light bulbs you have in your house

WHAT DID WE LEARN?

- What does incandescent mean? **Something is incandescent if it emits light or glows when a current is passed through it.**
- Why is it important that there be no oxygen inside a light bulb? **The oxygen would combine with the filament's atoms to cause the filament to burn up.**
- Who is credited with inventing the first incandescent light bulb? **Thomas Edison.**

TAKING IT FURTHER

- How might a light bulb be designed to keep any stray oxygen atoms from burning up the filament? **Some filaments are covered with a material that reacts with any stray oxygen atoms in the bulb the first time it gets warm. This uses up any oxygen atoms and protects the filament.**
- Premium light bulbs are filled with krypton instead of argon or nitrogen. Krypton is a heavier element than the other inert gases. How might krypton make the light bulb better? **The heavier krypton atoms are more likely to push the tungsten atoms back to the filament instead of letting them float away. Therefore, the filament will last longer.**
- Some incandescent light bulbs claim to burn cooler than ordinary bulbs. Is this an advantage or disadvantage? **This is likely to be a disadvantage. The hotter the bulb burns, the more light it gives off. So a cooler burning light bulb will be dimmer and less energy efficient; however, if the bulb is being used in a location where it is likely to be touched, the heat given off might be a consideration.**
- Why do refrigerator light bulbs last so long? **A light bulb generally burns up because the filament becomes thin. This thinning occurs in direct proportion to the length of time the filament is hot. A refrigerator light bulb is on for only a short period of time each day, so the filament does not burn out very quickly.**

CHALLENGE: ALTERNATIVE BULBS

- **Halogen bulbs are brighter and whiter than incandescent bulbs so can be used when a brighter light is needed, such as in headlights for an automobile; also, they last longer so might be used in areas that are hard to reach, or where it is difficult to replace a bulb.**
- **Although a fluorescent bulb is more expensive initially, it lasts so much longer and uses so much less power to produce the same amount of light that in the long run you will save money.**
- **LEDs use even less energy and last even longer than fluorescent bulbs so they are rapidly replacing incandescent and fluorescent bulbs in many applications.**

Refrigeration

Keeping things cool

Supply list

Rubbing alcohol	Water	Washcloth	Ice cubes	Rubber band

What did we learn?

- What is refrigeration? **The removal of heat.**
- What is the ideal temperature for the inside of a refrigerator? **35–38°F (1.7–3.3°C)**
- Why is this the ideal temperature range? **Any colder and the food would freeze, any warmer and the food would spoil too quickly.**
- What substance is used as a refrigerant in most new refrigerators? **HFCs—hydrofluorocarbons.**

Taking it further

- Can you cool your kitchen by leaving the refrigerator door open? Why or why not? **No. The heat from the room will enter the refrigerator where it will be absorbed by the refrigerant. But that heat will be released back into the room as the refrigerant is compressed and condenses so the room will not become cooler. Your food will become warmer though.**
- What is the best way to make your refrigerator more efficient? **Keep the door closed.**
- When you open the refrigerator door, some moisture from the air condenses on the evaporation, or expansion, coils. How might you stop a build up of ice from becoming a problem? **There are probably many solutions, but today's frost-free refrigerators have a small heating coil attached to the evaporation coils. This heating coil comes on every few hours and melts the ice that has accumulated. This water drips down to a pan underneath the refrigerator where it evaporates back into the air.**
- How does a refrigerator keep the inside from becoming too cold? **There is a thermostat inside your refrigerator that controls how often the compressor comes on so that the inside temperature is maintained within an acceptable range.**

Challenge: Chemical Refrigerators

- **RVs are smaller than houses and generally do not have room for a regular refrigerator. Also, electricity is not always available, so a process that uses a portable energy source is preferable and RVs generally have propane available at all times.**
- **Chemical refrigerators are similar to mechanical refrigerators because they both have substances that change state.**
- **They are different because one uses a mechanical compressor and the other uses chemical reactions to bring about this change of state.**

Sewing Machine

The treadle

Supply list

Sewing machine   Thread   Sewing needle   Cloth

What did we learn?

- What are some of the problems that had to be solved before a useful sewing machine could be built? **There had to be a way to make stitches without passing the entire needle completely through the fabric; the fabric had to be moved through the machine; the fabric had to be held in place while the stitch was being made; there had to be a way to regulate the thread through the machine.**
- What were some improvements made by Isaac Singer? **He designed a way to move the fabric horizontally; he made a presser foot to hold the fabric in place; he designed a wheel to move the fabric; he designed a foot treadle to free up the seamstresses' hands.**

Taking it further

- How is sewing with a machine similar to sewing by hand? **A needle is used to take thread through a piece of cloth and make stitches.**
- How is sewing with a machine different from sewing by hand? **In a machine, the eye of the needle is at the bottom of the needle so the entire needle does not need to pass completely through the fabric. Two threads are used instead of one to make stitches. Machines are much faster and more precise than hand sewing.**
- How did the Civil War affect the demand for sewing machines? **The need for uniforms greatly increased the demand for sewing machines because the uniforms could be made more quickly.**

Modern Appliances

What's in your house?

Supply list

Plastic bag   Net bag   String   Measuring cup   Washcloth
Supplies for Challenge: Copy of "Household Appliances" worksheet

What did we learn?

- What is the purpose of an electrical appliance? **To save time and effort.**
- What are some of the major appliances in your home? **Answers will vary.**
- Explain how one of these devices saves you time and effort. **Answers will vary.**

Taking it further

- If you were to invent a new appliance what would it be and how would it work? **Answers will vary.**

Challenge: Time-savers

- **Answers will vary**.

Clocks

What time is it?

Supply list

String Stopwatch Metal nut or washer Tape

Supplies for Challenge: 2-liter plastic bottle Soda straw Modeling clay Masking tape Bucket or large container 2 half-gallon milk cartons Copy of "Water Clocks" worksheet

What did we learn?

- What is a clock? **A device to help us tell what time it is or how much time has passed.**
- What is the main difference between a mechanical clock and an electric clock? **The power source—a mechanical clock has a spring or other device that is mechanically wound to provide power. An electric clock uses a battery or other source of electricity to provide power.**
- Why was a pendulum an important part of an accurate clock? **The pendulum swings at the same speed regardless of how high it is swinging so it keeps swinging at the same rate even when it has less energy in the swings.**

Taking it further

- Why was an accurate clock so necessary for navigation? **Sailors needed a way to determine how far they had traveled east or west, and the stars do not provide an east or west marker like the north star does.**
- Other than navigation, name two areas where accurate clocks are needed. **For scheduling transportation such as airplanes, trains, or buses. Also, competitions such as the Olympics or other athletic events need accurate clocks. In science, timing chemical reactions can be very critical. Nearly anything that has a computer in it needs an accurate clock.**

Challenge: Water Clocks worksheet

- Did the cartons empty at the same rate? **No.**
- Why do you think one carton emptired faster than the other one? **The carton held vertically has more water pressure—a taller column of water, so it empties faster.**
- Does the water flow from the bottle at a constant rate? Why/why not? **Water flows faster when the bottle is full; there is more water pressure, so marks will not be even.**
- How could you change the design of the clock in order to make the marks evenly spaced? **Change the shape of the container—perhaps a funnel shape.**
- What are some problems with a water clock? **Someone has to continuously fill the clock, and it only tells elapsed time not the actual time.**

Modern Conveniences

Lessons 22–26

Fill in the blank with the correct term from below.

1. Thomas Edison invented the _**incandescent**_ light bulb.
2. Premium light bulbs are filled with _**krypton**_ gas.
3. All _**oxygen**_ gas must be removed from a light bulb to prevent burning of the filament.
4. _**Refrigeration**_ is the removal of heat.
5. _**Food preservation**_ is the main reason refrigerators were invented.
6. A _**presser foot**_ holds fabric in place on a sewing machine.
7. A _**shuttle**_ is used on both a sewing machine and a loom to carry one thread through another.
8. Modern appliances are designed to save _**time**_ and _**effort**_.
9. A mechanical clock uses a _**pendulum**_ to regulate time.
10. _**Cesium**_ is the element used to regulate time in an atomic clock.
11. Clocks were instrumental in improving _**navigation**_ for sailors.
12. Mechanical clocks are often powered by a _**spring**_.
13. Electronic clocks are often powered by a _**battery**_.
14. In a washing machine, _**centripetal**_ force helps remove water from clothes.
15. Refrigeration requires a substance to change _**state**_.

Challenge Questions

Mark each statement as either True or False.

16. _**F**_ Halogen light bulbs are less efficient than regular incandescent bulbs.
17. _**T**_ Fluorescent bulbs create ultraviolet light with causes phosphors to glow.
18. _**T**_ LEDs use a semiconductor to produce light.
19. _**F**_ Chemical refrigerators do not use condensation.
20. _**T**_ Chemical refrigerators use heat to cause the refrigerant to change state.
21. _**T**_ Sewing machines use a shuttle to connect two threads together.
22. _**F**_ Modern appliances do not save us much time.
23. _**F**_ Water in a water clock always flows at a steady rate.
24. _**F**_ Water clocks are a relatively modern invention.

UNIT 5
MEDICAL INVENTIONS

MICROSCOPE

OPENING UP A WHOLE NEW WORLD

SUPPLY LIST

Copy of "Parts of a Microscope" worksheet

PARTS OF A MICROSCOPE WORKSHEET

1. **Eyepiece**
2. **Coarse adjustment**
3. **Fine adjustment**
4. **Objective**
5. **Stage**
6. **Clips**
7. **Light source**
8. **Base**

WHAT DID WE LEARN?

- What is a microscope? **A device that magnifies very small items so they can be viewed.**
- How is a simple microscope different from a compound microscope? **The simple microscope has one lens and the compound microscope has two lenses. The second lens magnifies the magnified image.**
- How did the achromatic lens improve the microscope? **It eliminated much of the distortion caused by earlier lenses.**
- Name three simple medical inventions that are still used today. **Stethoscope, thermometer, and syringe.**

TAKING IT FURTHER

- If you wanted to view viruses, would you use an optical or electron microscope? **Viruses are many times smaller than bacteria and must be viewed with an electron microscope.**
- When might a simple microscope be better than a compound microscope? **If your lenses are not very good on the compound microscope, the second lens will just magnify the distortion, producing a bad image. A smaller clearer image may be more useful than a larger distorted image. Van Leeuwenhoek found this to be true. Even though some compound microscopes were being experimented with, their lenses were so bad that he decided a single good quality lens was better, so he made his own lenses.**

Medical Imaging—Part 1

X-rays and CT Scans

Supply list

Orange Knife

What did we learn?

- How does an X-ray work? **A beam of energy passes through a person. Some substances absorb the X-rays and others allow the X-rays to pass through. The X-rays that pass through are detected by a special photographic plate or other material and an image is formed of the inside of the person. Note that radiographs have traditionally been made on a special film; however, new digital technology now makes film unnecessary in many applications.**
- What are some common uses of a radiograph X-ray? **To view bones, to check for breaks, and to look for cavities in teeth.**
- What is fluoroscopy used for? **To view organs in motion.**
- What is a CT scan used for? **To generate a 3-D image of soft tissue or organs.**

Taking it further

- How can you make a 3-dimensional image from 2-dimensional pictures? **If many pictures are taken in evenly spaced slices, the slices can be put together to form the 3-D image with the computer filling in any missing details.**
- Which type of X-ray technology is best for viewing solid objects? **Radiography.**
- How can X-ray technology be useful in a war hospital? **Not only can X-rays help identify broken bones, but they can also be used to locate shrapnel, bullets, and other foreign articles inside a body.**
- How might X-ray technology be useful to archeologists? **One important use of X-rays is for viewing inside mummies without destroying them. Scientists can not only view the skeleton of the person, but see amulets and other artifacts that were included with the mummy.**

Challenge: X-rays in Industry

- **X-rays show shape and density of items in a suitcase. Operators are trained to look for certain shapes. An X-ray of a soda can reveals the level of soda in the can because the soda absorbs a small amount of the X-ray energy. If the soda level is too low, the can is removed from the assembly line.**

Medical Imaging—Part 2

Ultrasound and MRI

Supply list

2 Magnets Compass

What did we learn?

- What are two forms of medical imaging that do not use X-rays? **Ultrasound and MRI.**
- How does an ultrasound machine generate images? **By passing sound waves into the body and using the echoes of those sounds to generate an image.**
- How does an MRI generate an image? **A magnetic field is used to orient atoms in the body; then radio waves disrupt the atoms. The machine measures the time for these atoms to realign and uses that information to generate an image.**
- Why are ultrasound and MRI preferable to X-rays in some instances? **X-rays can damage cells and would not be safe in some instances. Also, an MRI can sometimes detect conditions that X-rays cannot detect.**

Taking it further

- Why must a person undergoing an MRI remove all metal from her body? **The metal would affect the magnetic field.**
- The newest ultrasound is sometimes referred to as 4-D ultrasound. Why do you think it is called this? **With the development of faster computers, three dimensional images of babies in the womb can be seen as they are moving in nearly real time. As computer technology improves, ultrasound will show the baby in three dimensions as it is actually moving, allowing parents and doctors to view the child in very realistic images before it is born.**

Challenge: Doppler Ultrasound

- **Just as the frequency of the sound waves change as the object approaches, so the ultrasound wave frequency also changes as the blood moves toward or away from the transducer. Thus, the frequency difference can be used to indicate the direction and speed of the blood flow.**

Microsurgery

Keeping it small

Supply list

2 oranges (not seedless) Tape Knife Cutting board Tweezers

What did we learn?

- What is microsurgery? **Surgery that involves small incisions instead of large ones.**
- List at least three inventions that have made microsurgery possible? **There are many, but some listed in this lesson include microscope, miniature camera/endoscope, computer, robotic arms, and joystick controls.**
- Why is microsurgery often better than traditional surgery? **It causes less trauma to the patient.**
- Name three advantages to using robotic arms for microsurgery? **They are more precise; they filter out hand tremors; they cause less trauma to the patient; they are more cost effective.**
- What are three parts of a robotic surgical system? **Robot arms, computer workstation, video system, and joystick controls.**

Taking it further

- What is one disadvantage to robotic surgery? **One disadvantage is that the doctor cannot actually feel the tissues. However, special probes are being developed that would allow a doctor to use the robotic arm to remotely "feel" the tissues.**
- What might be some advantages to tele-surgery? **The patient would not have to travel to a specialist to have a special procedure done. This saves time and money. Also, astronauts could have surgery while still in space, or people in remote areas like research outposts in Antarctica could have surgery without having to leave. Can you think of others?**

Challenge: Designing a Robotic Surgery System

- **Answers will vary.**

Medical Inventions

Lessons 27–30

Explain how each of the following inventions has helped improve medical care.
Accept any reasonable answers.

1. Microscope: **Allows doctors to see bacteria and cells.**
2. X-ray: **Allows doctors to see broken bones and solid objects inside the body.**
3. CT scan: **Allows doctors to see soft tissues.**
4. Ultrasound: **Allows doctors to see developing babies in the womb and to see other parts of the body without using X-rays.**
5. MRI: **Allows 3D images to be made of nearly any part of the body.**
6. Robotics: **Allows doctors to perform microsurgery.**
7. Thermometer: **Shows elevated temperature to indicate possible illness.**
8. Stethoscope: **Amplifies heartbeat and other sounds.**
9. Endoscope: **Allows doctors to see inside the patient without major surgery.**

Challenge questions

Short answer:

10. How are electron microscopes different from optical microscopes? **Electron microscopes use electromagnetic coils and streams of electrons to magnify an image instead of light and lenses.**
11. What is one non-medical application for X-rays? **Checking for micro-cracks in metal, scanning for weapons at airports, or quality control.**
12. How is Doppler ultrasound different from regular ultrasound? **Doppler ultrasound uses the change in frequency of the sound waves to not only detect objects but to tell how they are moving inside the body.**

Unit 6
Entertainment

Lesson 31
Roller Coasters
Ahhhh!

Supply list

Clear, flexible tubing (able to fit marble or BB)
Marble or BB
Tape and other materials to support tubing

What did we learn?

- What are the physical laws governing the operation of a roller coaster? **Gravity and Newton's three laws of motion.**
- What are the names of the three wheels on roller coaster cars and the purpose of each? **The top wheel, or road wheel, guides the car along the track. The bottom wheel, or upstop wheel, keeps the car from slipping vertically. The sideways wheel, or guide wheel, keeps the car from slipping sideways.**
- What are some safety features that are built into roller coasters? **The three wheel system, restraints, computer sensors and controls, brakes, extensive testing.**
- Explain why a roller coaster car does not stop at the bottom of the first hill. **The car has momentum and inertia. Remember, an object in motion tends to stay in motion according the Newton's First Law of Motion. Thus, unless brakes are applied, there is probably not enough friction to stop the car at the bottom of the hill, and its momentum will carry it up the next hill.**

Taking it further

- What would happen if the first hill was not the highest hill on the track? **Unless there is some other method provided for getting the car up the highest hill, the car would not have enough energy/momentum to make it to the top of a higher hill, and it would eventually come to rest at the bottom of the hill.**
- Why doesn't the car have enough energy to make it to the top of a higher hill, or even one the same height, as the first one? **Some of the energy is continually being lost to friction and wind resistance.**
- What are the sources of friction in a roller coaster? **Friction between the wheels and the track, between the cars and the air molecules, and between the passengers and the air molecules.**

Phonograph

Listen to that sound

Supply list

Aluminum foil   Scissors   Cardboard   Flashlight or laser pen

What did we learn?

- What is a phonograph? **A device that could record and play back sounds.**
- Who invented the first phonograph? **Thomas Edison.**
- Which devices were later developed from the idea of the phonograph? **The gramophone, dictaphone, record players, tape players, CD players, and MP3 players, to name a few.**
- What recording technology uses electromagnets? **Magnetic tapes.**

Taking it further

- What are some advantages of CDs over vinyl records? **The digital signal is clearer so you get better sound. The laser reader does not wear the CD out like a needle does to a vinyl record. CDs are smaller and more portable.**
- List at least three devices that use the idea of changing sound into a different form, then changing it back again. **Telephone, CD recorder/player, tape recorder, DVD recorder, computer, Sonar, and video camera.**
- Other than entertainment, what are some uses for recording and playing back sound? **Communications, research, education, dictation, etc.**

Challenge: MP3 players

- **An MP3 player has a data port to talk to a computer, memory to store the information, microprocessor to control the flow of information, digital signal processor to convert the digital information into an analog signal, amplifier to make the sound louder, audio port for the analog signal to come out (which must be connected to earphones or some other speaker), display and playback controls to allow you to talk to the player, and a power supply.**

Moving Pictures

You ought to be in pictures!

Supply list

Multiple copies of the "Movie Frames" worksheet   Pen or pencil   Scissors   Stapler

What did we learn?

- What is a motion picture? **A series of still photos that are flashed in front of your eyes quickly enough that your mind blends them together.**
- Who were some of the first people to make moving pictures? **Thomas Edison and the Lumière brothers.**
- What is persistence of vision? **The image of an object can still be seen by your brain for a short period of time after the image is removed from your sight.**

Taking it further

- How did the phonograph influence the development of movies? **First, it gave Edison the idea to do for the eye what the phonograph did for the ear. Second, it provided sound for the movies.**
- Why did the Black Maria have a roof that could be opened? **To allow sunlight to illuminate the stage.**

Becoming an Inventor

You could be the next Thomas Edison.

Final Project supply list

Sketch paper Various materials to make the invention

Supplies for Challenge: Research materials on chosen invention

What did we learn?

- What are the steps to inventing something? **Come up with an idea, design the invention, build it, test it, and patent it.**
- How is the scientific method similar to the process for inventing something? **Both methods recognize a problem, suggest a solution, design a way to test the solution, collect data, and then publish the results.**
- How are the two processes different? **Mainly their goals can be different. The inventor wants to develop something new, but the scientist may only want to understand how something works.**

Taking it further

- What can you do to become a better inventor? **With permission, take things apart to see how they work, learn scientific principles, study nature, read about other inventions, learn to draw, and experiment.**

Entertainment

Lessons 31–34

Explain how a roller coaster makes use of each of the following. **Accept reasonable answers**

1. Gravity: **Gravity pulls down on the cars giving them the necessary momentum to continue up the next hill.**
2. Inertia: **The inertia of the car translates into momentum so the car does not stop until an outside force is applied.**
3. Newton's third law of motion: **When reaching the bottom of a hill or the top of a loop, the car pushes against the rider with the same force that the rider pushes against the car.**

Short answer:

4. Name four inventions that were designed to record and play back sound. **Phonograph, Gramophone, Dictaphone, Tape player, CD player, MP3 player.**
5. Name three different materials that have been used to record sound. **Metal foil, wax, vinyl, magnetic tape, reflective plastic disks.**
6. Explain how still pictures appear to be moving in motion pictures. **There are only slight differences in each still picture and they appear so close together that your mind blends them together and they appear to be moving.**
7. List the five steps to becoming an inventor.

   A. **Come up with an idea** B. **Design the invention** C. **Build it** D. **Test it** E. **Patent it**

Challenge questions

Short answer:

8. Describe some of the effects a roller coaster has on the human body. **Squeezes your stomach, confuses your bone and muscle receptors, causes you to feel dizzy.**
9. What is one advantage to using an MP3 player for listening to music? **Uses less memory, can get music from multiple sources, small and portable.**
10. Describe the claymation process. **Objects or characters are made from clay, photographed, and then moved to a slightly different position over and over again. Then the slightly different images are put together and shown very quickly.**

Inventions & Technology

Lessons 1–34

Match each term with its definition.

1. _**C**_ Detecting objects with radio waves
2. _**E**_ Detecting objects with sound waves
3. _**A**_ A copy of something
4. _**B**_ Cathode ray tube

5. _**D**_ Brain of the computer
6. _**G**_ Objects that can be discarded to lighten load
7. _**F**_ Random access memory
8. _**H**_ Liquid refrigerants

Mark each statement as either True or False.

9. _**F**_ Coming up with the idea for an invention is usually the hardest part.
10. _**F**_ Only very special people can be inventors.
11. _**T**_ Bell and Edison both contributed to the invention of the telephone.
12. _**T**_ Radio was developed about 50 years after the telegraph.
13. _**F**_ A TV signal contains only picture information.
14. _**T**_ Computers store information as a series of 1s and 0s.
15. _**T**_ Invention of the steam engine helped spur the Industrial Revolution.
16. _**T**_ Trains today primarily operate with diesel engines.
17. _**F**_ Diesel engines use spark plugs.
18. _**T**_ Henry Ford made the automobile affordable for many people.
19. _**T**_ Gunpowder revolutionized warfare and weaponry.
20. _**F**_ Electric light bulbs last longer if they are filled with oxygen.

Match the inventor with his invention.

21. _**C**_ Johann Gutenberg
22. _**B**_ James Watt
23. _**D**_ Samuel Morse
24. _**F**_ Guglielmo Marconi
25. _**A**_ Werner Von Braun
26. _**G**_ Ernest Swinton
27. _**I**_ John Phillip Holland
28. _**J**_ Elias Howe
29. _**H**_ Anton Van Leeuwenhoek
30. _**E**_ Wilhelm Roentgen
31. Name at least three things invented by Thomas Edison. **Electric light bulb, many items needed for an electric system/power company, talking doll, phonograph, kinetograph/moving pictures, improved telephone, duplex telegraph, improved ticker, the list goes on and on.**
32. The printing press was invented in about 1455, but most of the inventions we studied were invented in the last 300 years. Place the following inventions on the timeline in the order in which they were invented. Use the book to look up the date that each was invented.

1782—Steam engine **1837—Telegraph** **1840—Sewing machine** **1876—Telephone**
1877—Phonograph **1879—Light bulb** **1903—Airplane** **1927—Television**
1930—Jet engine **1933—Electron microscope** **1970—CT scan**

Challenge questions

Short answer:

33. Chose one invention not covered in this book and describe how you think it works. **Accept reasonable answers.**
34. Describe what you would do to become an inventor. **Accept reasonable answers.**

Conclusion

God made you creative

Supply list

Bible

Resource Guide

Many of the following titles are available from Answers in Genesis (www.AnswersBookstore.com).

Suggested Books

How to be an Inventor by Murray Suid—a highly-recommended guide to helping your child become an inventor

The Story of Inventions by Michael J. McHugh and Frank P. Bachman—a more detailed look at many of the inventions in this book as well as several not covered here

How Things Work, by Neil Ardley, from Reader's Digest—100 ways that parents and children can share the secrets of technology; fun with more advanced activities

Clocks—Building and Experimenting with Model Timepieces by Bernie Zubrowski—many interesting activities for clocks from water clocks to alarm clocks

Exploring the World of Physics by John Hudson Tiner—explains physics in clear detail, using ordinary speech

Exploring the World of Mathematics by John Hudson Tiner—explains mathematics from a fun historical approach

Science and the Bible Volumes 1–3 by Donald B. DeYoung—scientific demonstrations with biblical truths

Invention Mysteries and *More Invention Mysteries* by Paul Neumann—Fun "rest-of-the-story" style explanations of many inventions

Suggested Videos

Newton's Workshop by Moody Institute—Excellent Christian science series; several titles to choose from

Field Trip Ideas

- Visit the Creation Museum in Petersburg, Kentucky.
- Tour any factory in your town.
- Visit a history museum to see what inventions were like in that time period.
- Tour a hospital or doctor's office and ask to see the equipment they use.
- Tour a radio or TV station.
- Ride a roller coaster and observe its design.

Creation Science Resources

Answers Book for Kids Four volumes by Ken Ham with Cindy Malott—Answers children's frequently asked questions

The New Answers Books 1 & 2 by Ken Ham and others—Answers frequently asked questions

The Amazing Story of Creation by Duane T. Gish—Gives scientific evidence for the creation story

Creation Science by Felice Gerwitz and Jill Whitlock—Unit study focusing on creation

Creation: Facts of Life by Gary Parker—Comparison of the evidence for creation and evolution

The Young Earth by John D. Morris—Lots of facts disproving old-earth ideas

Master Supply List

The following table lists all the supplies used for *God's Design for the Physical World: Inventions & Technology* activities. You will need to look up the individual lessons in the student book to obtain the specific details for the individual activities (such as quantity, color, etc.). The letter *c* denotes that the lesson number refers to the challenge activity. Common supplies such as colored pencils, construction paper, markers, scissors, tape, etc., are not listed.

Supplies needed (see lessons for details)	**Lesson**
Aluminum foil	8, 32
Bag (mesh)	25
Bag (plastic)	25
Balloons	14
Battery (6-volt)	4
BBs (steel)	31
Bible	35
Bolts	10c
Bucket	26c
Cloth	24
Coins (pennies, nickels)	9
Colored filters or colored plastic wrap	3c, 5
Compass (navigational)	29
Craft sticks	9, 19
Cups (paper or foam)	3
File (metal)	4
Flashlights	3, 5, 32
Flour	18
Ink pad	1
Ketchup packets or other condiments packets	20
Magnets	9, 29
Magnifying glass	6c
Marbles	31
Milk carton (½-gallon)	26c
Modeling clay	3, 4, 16, 19, 26, 33
Needle (sewing)	24
Notebook	34
Nuts or washers	26

Supplies needed (see lessons for details)	**Lesson**
Oranges	28, 30
Oven mitts	8, 10
Paper clips	3
Plastic bottle (2-liter, empty)	20, 26
Plate (foam or paper)	16
Popcorn (unpopped)	18
Poster board/cardboard/tagboard	9, 10, 14, 16, 32
Potatoes	1
Radio or CD player	4
Rice (uncooked)	18
Rubber bands	23
Rubber stamps	1
Rubbing alcohol	23
Sewing machine	24
Stopwatch	26
Straws	9, 14, 16, 26
String	3, 4, 8, 14, 25, 26
Thread (spools)	19, 24
Toothpicks	16c
Train (toy)	9
Tubing (clear plastic)	3, 31
TV remote control (optional)	5
Tweezers	30
Wire (copper)	4
Wood (block)	17
Yard stick/meter stick and ruler	6c, 17

Works Cited

"Adding Firepower with the Invention of Gunpowder." http://www.dummies.com/WileyCDA/DummiesArticle/id-1225.html.

Adkins, Jan. *Bridges*. Brookfield: Roaring Book Press, 2002.

Adler, David A. *Thomas Alva Edison*. New York: Holiday House, 1990.

Alter, Judy. *Samuel F. B. Morse*. Chanhassen: The Child's World, 2003.

Ardley, Neil. *How Things Work*. Pleasantville: Reader's Digest, 1995.

Bellis, Mary. "History of the Fax Machine & Alexander Bain." http://inventors.about.com/library/inventors/blfax.htm.

Bender, Lionel. *Invention*. New York: Alfred A. Knopf, 1991.

Birch, Beverley, and Robin B. Corfield. *Marconi's Battle for Radio*. London: Barrons, 1995.

Bonsor, Kevin, et al. "How MP3 Players Work." http://www.howstuffworks.com/mp3-player.htm.

Bonsor, Kevin. "How Robotic Surgery Will Work." http://electronics.howstuffworks.com/robotic-surgery1.htm.

Brain, Marshall. "How Gas Turbine Engines Work." http://science.howstuffworks.com/turbine5.htm.

Brain, Marshall. "How Refrigerators Work." http://www.howstuffworks.com/refrigerator.htm.

Burch, Joann J. *Fine Print*. Minneapolis: Carolrhoda Books, 1991.

Carlisle, Norman, and Madelyn Carlisle. *Bridges*. Chicago: Childrens Press, 1983.

"Cassini-Huygens Mission to Saturn and Titan." http://saturn.jpl.nasa.gov/home/index.cfm.

"The Chariot." http://nefertiti.iwebland.com/timelines/topics/chariot.htm.

Ciovacco, Justine. *Turn on the TV*. San Diego: Blackbirch Press, 2004.

Coiley, John. *Train*. New York: Dorling Kindersley, 2000.

Cook, Nick. *Roller Coasters or I Had So Much Fun, I Almost Puked*. Minneapolis: Carolrhoda Books, Inc., 1998.

Delano, Marfe F. *Inventing the Future*. Washington D.C.: National Geographic, 2002.

"Cooler M1 Abrams Tank Engines." http://www.g2mil.com/abramsdiesel.htm.

Duffy, Trent. *The Clock*. New York: Atheneum Books for Young Readers, 2000.

"Fax Machine." http://www.ideafinder.com/history/inventions/fax.htm.

"The First True Incandescent Light Bulb." http://www.maxmon.com/1878ad.htm.

Fisher, Leonard E. *Gutenberg*. New York: Macmillan Publishing Co., 1993.

Fisher, Leonard E. *The Printers*. New York: Benchmark Books, 2000.

"Five Years on the International Space Station." http://www.nasa.gov/mission_pages/station/main/index.html.

Fox, Mary V. *Inventors & Inventions*. New York: Benchmark Books, 1996.

"Frederick M. Jones." http://www.eia.doe.gov/kids/history/people/pioneers.html#Jones.

"Fuel Cells." http://www.eere.energy.gov/hydrogenandfuelcells/fuelcells/basics.html.

Gearhart, Sarah. *The Telephone*. New York: Atheneum Books for Young Reader, 1999.

Griffith, Susan. "Dittrick exhibit looks at medical history through microscope." http://www.cwru.edu/pubs/cnews/2000/10-12/microscope.htm.

"Gunpowder." http://www.wavespring.com/justin/china/gunpowder.html.

"Henry Ford." http://inventors.about.com/library/inventors/blford.htm.

Herbert, Stephen. "Persistence of Vision." http://www.grand-illusions.com/articles/persistence_of_vision/.

"A History of the Computer: Electronics." http://www.pbs.org/nerds/timeline/elec.html.

"The History of the Motion Picture." http://inventors.about.com/library/inventors/blmotionpictures.htm.

Hockman, Hilary, Ed. *What's Inside Everyday Things.* New York: Dorling Kindersley, Inc., 1992.

Hockman, Hilary, Ed. *What's Inside Great Inventions.* New York: Dorling Kindersley, Inc., 1993.

"How Flintlock Guns Work." http://www.howstuffworks.com/flintlock.htm.

"Incandescent Light Bulbs." http://howthingswork.virginia.edu/incandescent_light_bulbs.html.

"International Space Station." http://www.nasa.gov/mission_pages/station/main.

Jefferis, David. *The First Flyers.* London: Franklin Watts, 1988.

"Jet Propulsion 101." http://www.geae.com/education/engines101.

"Jonas Salk M.D." http://www.lucidcafe.com/library/95oct/jesalk.html.

Kerby, Mona. *Samuel Morse.* New York: Franklin Watts, 1991.

Mallard, Neil. *Submarine.* London: Dorling Kinderserly, 2003.

Mine, Mark. "How Offset Printing Works." http://www.howstuffworks.com/offset-printing.htm.

Mulcahy, Robert. *Medical Technology: Inventing the Instruments.* Minneapolis: Oliver Press, Inc., 1997.

Naden, Corrine J., and Rose Blue. *Jonas Salk: Polio Pioneer.* Brookfield: Millbrook Press, 2001.

Nemes, Claire. *Young Thomas Edison.* Troll Associates, 1996.

"Persistence of Vision." http://www.exploratorium.edu/snacks/persistence_of_vision/index.html.

Petty, Kate. *I Didn't Know That Some Planes Hover.* Brookfield: Copper Beech books, 1998.

"Press Release: 1986 Nobel Prize Physics." http://nobelprize.org/physics/laureates/1986/press.html.

"Radar." Columbia Electronic Encyclopedia, http://www.encyclopedia.com/doc/1E1-radar.html.

Richie, Jason. *Weapons: Designing the Tools of War.* Minneapolis: Oliver Press, 2000.

Rutland, Jonathan. *The Age of Steam.* New York: Random House, 1987.

Sarro, Dave. "How a Refrigerator Works." http://wps.ablongman.com/long_gurak_cgtc_3/54/14046/3595904.cw/index.html.

Shaefer, A. R. *Roller Coasters.* Mankato: Capstone Press, 2005.

Siegel, Beatrice. *The Sewing Machine.* New York: Walker and Company, 1984.

Stwertka, Eve, and Albert Stwertka. *A Chilling Story How Things Cool Down.* Englewood Cliffs: Simon & Schuster, 1991.

Suid, Murray. *How to be an Inventor.* Palo Alto: Monday Morning Books, 1993.

Swanson, Gloria M., and Margaret V. Ott. *The Story of Frederick Mckinley Jones—I've Got an Idea!* Minneapolis: Runestone Press, 1994.

Taylor, Ron. *Journey Through Inventions.* New York: Smithmark, 1991.

Turnbull, Stephanie. *Trains.* London: Usborne Publishing, 2002.

Urquhart, David I. *The Internal Combustion Engine and How it Works.* New York: Henry Z. Walck, Inc., 1973.

"Weapons and Catapults." http://www.legionxxiv.org/weapons.

Weiss, Harvey. *Motors and Engines and How They Work.* New York: Thomas Y. Crowell Company, 1966.

Winkler, Kathy. *Radiology.* New York: Benchmark Books, 1996.

Zannos, Susan. *Godfrey Hounsfield and the Invention of CAT Scans.* Bear: Mitchell Lane Publishers, 2003.

Zubrowski, Bernie. *Clocks.* New York: Morrow Junior Books, 1988.